Nayara Melquiades de Oliveira
Antonio A. S. Alfaya

Silicon and Titanium Nanocomposites

Nayara Melquiades de Oliveira
Antonio A. S. Alfaya

Silicon and Titanium Nanocomposites

Synthesis, characterisation and application as catalysts in the degradation of Methylene Blue using sunlight

ScienciaScripts

Imprint
Any brand names and product names mentioned in this book are subject to trademark, brand or patent protection and are trademarks or registered trademarks of their respective holders. The use of brand names, product names, common names, trade names, product descriptions etc. even without a particular marking in this work is in no way to be construed to mean that such names may be regarded as unrestricted in respect of trademark and brand protection legislation and could thus be used by anyone.

Cover image: www.ingimage.com

This book is a translation from the original published under ISBN 978-3-330-99596-3.

Publisher:
Sciencia Scripts
is a trademark of
Dodo Books Indian Ocean Ltd. and OmniScriptum S.R.L publishing group

120 High Road, East Finchley, London, N2 9ED, United Kingdom
Str. Armeneasca 28/1, office 1, Chisinau MD-2012, Republic of Moldova, Europe
Managing Directors: Ieva Konstantinova, Victoria Ursu
info@omniscriptum.com

Printed at: see last page
ISBN: 978-620-8-58760-4

CONTENTS

CHAPTER 1

INTRODUCTION

The textile industry consumes a large amount of water during the manufacture, dyeing and finishing of its products. The loss of dye during these processes leads to various problems, such as the pollution of rivers and lakes, leading to eutrophication (GALINDO et al. 2001; HOUAS et al. 2001; BIANCO-PREVOT et al. 2001). The Colour Index (Catalogue of the Society of Dyers and Colourists) currently lists more than 8,000 synthetic organic dyes associated with the textile industry. This diversity is justified by the commercial success of textile products that undergo a dyeing process. In order to meet an increasingly demanding market, the industry has invested in the development of economical dyes with specific properties to achieve good colour fixation in fabrics (ZANONI; CARNEIRO, 2001).

Dyes are molecules that comprise two key components: the chromophore group, responsible for colouring, and the functional group, which binds to the fabric fibres. The chromophore group most commonly used in the textile industry is the -N=N- functional group attached to aromatic systems. Dyes can have one or more of these functional groups per dye molecule, forming a family of dyes called azo dyes.

There are hundreds of known dyes in the literature, which are classified by their chemical nature or in terms of their application to the type of fibre: acid, direct, basic, sulphur and reactive. Basic (or cationic) dyes have a positively charged chromophore group (SOARES, 1998; KUNZ et al., 2002).

Methylene blue (Figure 1) is a basic dye belonging to the phenothiazine class. It is an organic, aromatic, heterocyclic compound that is soluble in water and alcohol. It has the molecular formula $C_{16}H_{18}ClN_3S.\ H_2O$ and a molar mass of 373.9 g mol^{-1}. This dye has various applications, such as in cheetah printing, dyeing, cotton printing and leather dyeing (WANG et al., 2008). It is also commonly used in the production of paper and other materials such as polyesters and nylon.

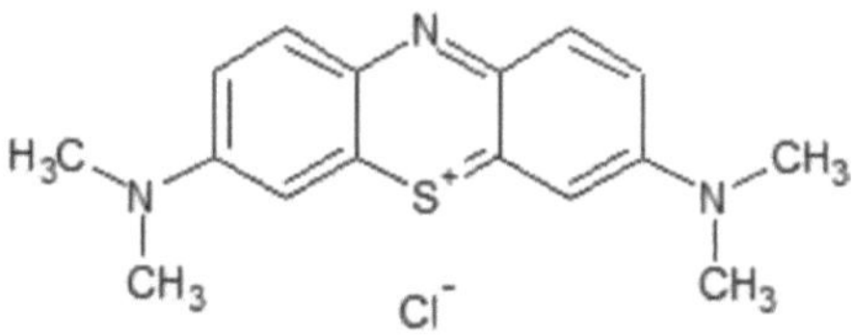

Figure 1 - Structure of the basic dye methylene blue

Source: Prepared by the author

The removal of dyes from effluents has been widely discussed in the textile sector due to their high biological stability to light and oxidising agents and, above all, because dyes do not belong to the same class of chemical compounds and therefore require specific methods for identification, quantification and degradation (ZANONI; CARNEIRO, 2001). Conventional systems used by industries, such as membrane filtration, solvent extraction, coagulation, among others, have not been very successful and are still costly and generate by-products that have an impact on the environment. For these reasons, new processes are continually being studied and their industrial performance evaluated (OZCAN et al., 2006; ROBINSON et al., 2001).

Advanced oxidation processes (AOPs) have been used as an alternative because they are destructive for most organic pollutants and are sustainable in the long term. These processes are based on the formation of hydroxyl radicals (HO"), a highly oxidising agent. Due to their high reactivity (Eo=2.8 V), these radicals can react with a wide variety of compound classes, promoting their total mineralisation into innocuous compounds such as CO_2 and water.

These processes are divided into homogeneous and heterogeneous systems, where hydroxyl radicals are generated with or without ultraviolet irradiation. These include processes involving the use of ozone, hydrogen peroxide and semiconductors such as titanium dioxide (heterogeneous photocatalysis) (NOGUEIRA; JARDIM, 1998).

The degradation of dyes through heterogeneous photocatalysis, a process that uses photons to trigger reactions catalysed by solid semiconductors, has been well documented. The principle of heterogeneous photocatalysis involves the activation of a semiconductor by sunlight or artificial light.

A semiconductor is characterised by valence bands (BV) and conduction bands (BC) and the region between them is called the "band-gap" (NOGUEIRA; JARDIM, 1998).

As shown in Equation 1, when a semiconductor is exposed to ultraviolet light, it is promoted to an electronically excited state characterised by the generation of an electron/laccase pair (e^-, h^+). The oxidising characteristics of the hydroxyl radical generated by the laccase reaction (Equations 2 and 3), and the strongly oxidising character of the laccase itself, mean that organic molecules adsorbed on the surface of the catalyst particle can be oxidised to the point of complete mineralisation through a very viable process (KUNZ et al., 2002).

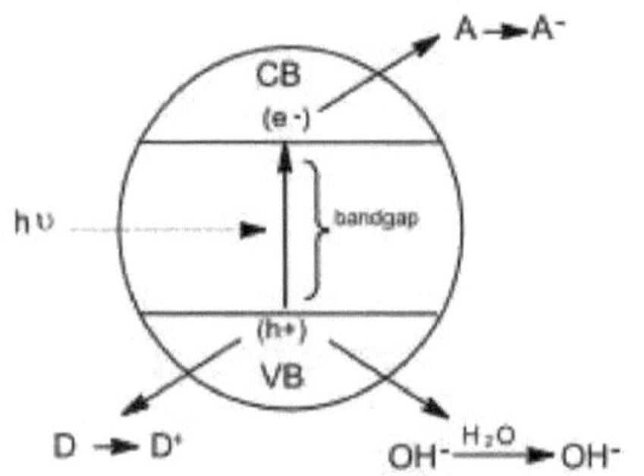

Figure 2 - Representation of the formation of the electron/lacuna pair (e^-, h).$^+$

Source: KUNZ et al.

$$TiO_2 + h\nu \rightarrow TiO_2\,(e^-_{(BC)}h^+_{(BV)}) \qquad \text{(Equação 1)}$$

$$H_2O + h^+ \rightarrow HO\bullet + H^+ \qquad \text{(Equação 2)}$$

$$OH^- + h^+ \rightarrow HO\bullet \qquad \text{(Equação 3)}$$

The efficiency of photocatalysis depends on the competition between the process in which the electron is removed from the surface of the semiconductor and the process of recombination of the electron/lacuna pair, which results in the release of heat, i.e. the species can recombine or migrate to the surface of the catalyst, where they can react with pre-adsorbed electron donor or receptor species, leading to redox reactions. The lower the recombination, the greater the efficiency of the semiconductor (NUNES, 2011).

Among the semiconductors, TiO_2 is the most widely studied due mainly to its non-toxicity, photostability and chemical stability over a wide pH range. However, the use of TiO_2 is not economically interesting for large-scale application, which forces research into suitable alternatives (BANSAL; SUD, 2011).

Hybrid materials and mixed oxides have attracted attention as heterogeneous catalysts and catalyst supports. These materials have one of the materials as the matrix, in which the particles of the second material are dispersed. The incorporation of inorganic fillers into polymers gives rise to materials with greater mechanical strength, greater thermal stability or superior optical, magnetic or electrical properties.

Fillers with nanometre dimensions (1-500 nm) have a high surface area, promoting better dispersion in the polymer matrix and, therefore, an improvement in the physical properties of the composite, which depend on the homogeneity of the material (ESTEVES et al., 2004). In addition to the large specific surface area, nanostructured photocatalysts

also present short migration distances for the photogenerated electron/lacuna pairs and a reduction in the possibility of recombination of the electron/lacuna pairs, preventing a decrease in photocatalytic activity.

Matrices of the SiO_2/MO_2 type (where M is a transition metal) can be obtained through sol-gel processes. This process consists of a low-temperature method for producing inorganic materials by hydrolysis and polycondensation reactions. These materials have physical rigidity, chemical inertia and thermal stability.

The addition of titanium to a silica matrix promotes interactions at the atomic level, which leads to the appearance of acid sites on the surface. These sites are not generally observed on pure oxides. The high concentration of titanium in mixed oxides can promote the appearance of interesting ion-exchange and structural properties (PEREIRA et al., 2002).

In this work, three matrices containing different concentrations of titanium were synthesised by the sol-gel process, using tetraethylorthosilicate (TEOS) and tetrabutylortho-titanate (TBOT) alkoxides as precursors. The matrices were characterised by chemical analysis, SEM, EDS, FT-IR, X-ray diffraction, diffuse reflectance and surface area (ASAP).

The aim of this work was to prepare SiO2/TiO2 nanocomposites with a high surface area and a high dispersion of titanium in the matrix so that they could be used to study the photodegradation of textile dyes using natural sunlight.

CHAPTER 2

EXPERIMENTAL PART

Reagents

For the preparation of the nanocomposites and photodegradation studies, the following reagents were used:

Tetraethylorthosilicate P.A. (Acros Organics),

Tetrabutyl orthotitanate P.A. (Aldrich),

Absolute ethanol (Kinetics),

Concentrated nitric acid (BIOTEC)

Methylene blue (Basic Blue 9, C.I. 52015 - Synth)

Milli-Qplus water.

Preparation of nanocomposites

Silicon and titanium nanocomposites were prepared in 80:20, 70:30 and 55:45 proportions, called ST20, ST30 and ST45, respectively. The system used to prepare the materials is shown in Figure 3.

In the "1^a hydrolysis" step, ethanol, TEOS, water and nitric acid were added in the quantities specified in Tables 1, 2 and 3. The mixture was kept stirring for two hours in a silicone bath at 80°C. The mixture was then cooled to 25°C.

In the second stage, known as the composite stage, TBOT and absolute ethanol were added to the mixture (Tables 1, 2 and 3). The system was kept stirring for two hours at 25°C.

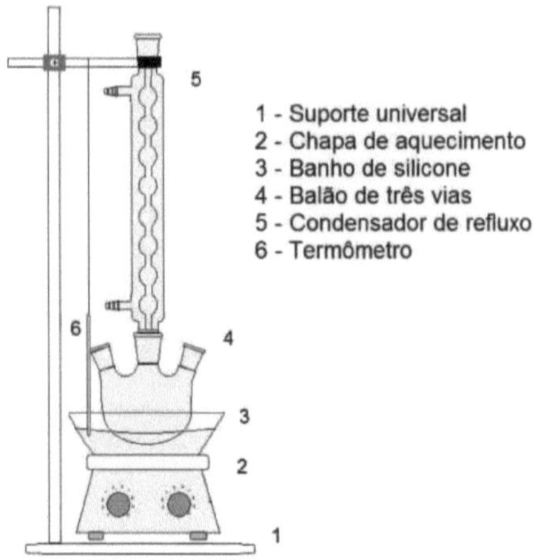

Figure 3 - Schematic of the apparatus used in the synthesis.

Source: Prepared by the author

In the third stage, called "2ª hydrolysis", nitric acid and water were added to the mixture in the quantities shown in Tables 1, 2 and 3. The mixture was kept stirring for two hours at 25°C and then transferred to a beaker and kept at 60°C until the system was completely gelatinised (Figure 4). The gel was kept in an oven at 110°C for 24 hours to obtain the xerogel.

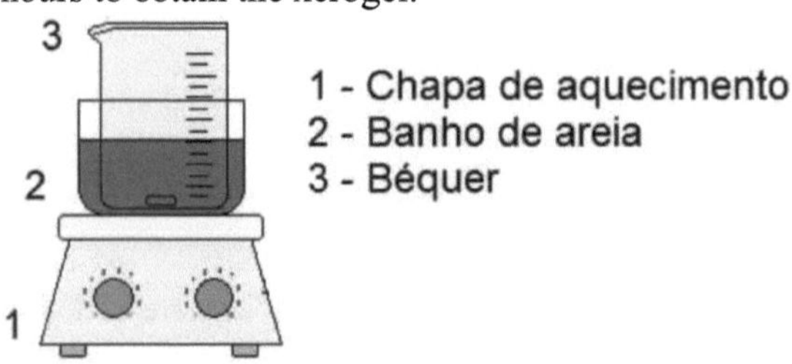

Figure 4 - Diagram of the gelatinisation system.

Source: Prepared by the author

Table 1- Quantities used in the synthesis of ST20.

Reagent	Quantity (mL)	Proportion in relation to TEOS
	1ª Hydrolysis	
TEOS	62,0	1
Ethanol	62,0	3,71
H2O	5,7	1,15

HNO3	0,4	0,018
	Composite	
TBOT	18,7	0,2
Ethanol	144,1	8,56
	2nd Hydrolysis	
HNO3	0,7	0,044
H2O	18,8	3,84

Table 2 - Quantities used in the synthesis of ST30.

Reagent	Quantity (mL)	Proportion in relation to TEOS
	1a Hydrolysis	
TEOS	62,0	1
Ethanol	62,0	3,71
H2O	5,7	1,15
HNO3	0,4	0,018
	Composite	
TBOT	28	0,3
Ethanol	216,1	12,84
	2nd Hydrolysis	
HNO3	1,1	0,066
H2O	28,2	5,76

Table 3 - Quantities used in the synthesis of ST45.

Reagent	Quantity (mL)	Proportion in relation to TEOS
	1ª Hydrolysis	
TEOS	62,0	1
Ethanol	62,0	3,71
H2O	5,7	1,15
HNO3	0,4	0,018
	Composite	
TBOT	42,0	0,45
Ethanol	324,3	19,46
	2nd Hydrolysis	
HNO3	1,6	0,099
H2O	42,7	8,635

The xerogel samples were ground in a knife mill and sieved to obtain particles with diameters between 0.15 and 0.25 mm. The materials in the desired particle size were calcined at 500°C for 120 hours with air flow.

Characterisation

Gravimetric analysis

To obtain the TiO2 content in the nanocomposites, the following procedure was used: 0.500 g of the nanocomposite under study and 1 or 2 drops of concentrated HF solution (40%) were added to a Teflon crucible. The mixture was left at room temperature for

5 minutes to completely solubilise the matrix. The solution was quantitatively transferred to a plastic beaker and diluted to 150 mL with Milli-Q water. The resulting solution was heated at 80 °C for 3 hours so that the silicon fluoride (SiF_4) formed was completely eliminated from the solution. The solution was then cooled to 25°C and 15 mL of concentrated ammonium hydroxide (NH_4OH) was slowly added. The solution was kept at 25°C for 24 hours to completely precipitate the hydrated TiO2. The mixture was quantitatively filtered through Whartmann 42 filter paper and the solid material was calcined at 900°C for 12 hours. The calcined material was weighed as TiO2 and the SiO2 content was assumed to be the remaining mass of the material.

Scanning electron microscopy (SEM)

SEM images and X-ray energy dispersive spectroscopy (EDS) mapping were carried out on a Phillips FEI Quanta 200 scanning electron microscope. The samples were deposited on carbon tape and metallised with gold.

Infrared spectrometry (FTIR)

For the infrared analyses, pellets were produced with KBr (1% by mass). The readings were taken on a Shimadzu FTIR-8300 infrared spectrometer with 240 accumulations.

X-ray diffraction

The powder X-ray diffraction analysis of the materials was carried out on an XPERT PRO MPD X-ray diffractometer, at 5.0250°< θ < *69.9750*, with an angular step of 0.05°, V=30 kV and *i=20 mA.*

Surface area and pore volume

The surface area and pore volume of the samples were determined using Quantachrome Autosorb1 (ASAP) equipment and the isotherm data was processed using the BJH mathematical method.

Diffuse reflectance

The diffuse reflectance spectra were obtained in a homemade system using an OceanOptics USB200+ portable spectrophotometer with integrated Spectra Suite software. The data was applied to a double flow model (Kubelka-Munk) using the Tauc graph to determine the gap band and to the effective mass model to determine the crystallite size.

Calculating the gap energy allows us to assess whether the nanocomposites can be effectively used as photocatalysts. To do this, the electron must cross the gap value, generating the electron/lacuna pair. The lower the energy, the greater the photocatalytic potential, since less energy is required for activation (NUNES, 2011).

To determine the gap energy, the approximate two-flux model (Kubelka-Munk) was used to investigate the absorption and scattering of radiation on the surface of the semiconductor. For samples produced by the sol-gel method, the surface of the coating is optically rough, and it is reasonable to use the Kubelka Munk model (MURPHY, 2007).

In the Tauc model, the gap energy is associated with the absorbance and energy of the photons. Equation 4 can be used to calculate the photon energy in eV (STENGL; KRÁLOVÁ, 2011).

$$E = \frac{hc}{\lambda} = \frac{1239{,}84}{\lambda} \qquad \text{(Equação 4)}$$

where E is the photon energy (eV) and λ *is the* wavelength (nm).

The absorption coefficient (a) is calculated using the Kubelka-Munk equation (Equation 5):

$$\alpha = \frac{(1-R)^2}{2R} \qquad \text{(Equação 5)}$$

where R is the reflectance.

To determine the gap value, the energy is multiplied by the value of the coefficient, making the absorbance quantised (Equation 6):

$$(\alpha E)^n = B^n E \qquad \text{(Equação 6)}$$

in which:

E is the energy in eV;

B^n represents the coefficient that expresses the degree of disorder at the sample band boundary (expressed in (eV - *cm-1*$^{)(n))}$;

n can take on different values depending on the type of transition between the bands. In this case, we consider an indirect transition and therefore *n* = *%* (MURPHY, 2007).

By carrying out the above calculations, data is obtained for constructing the Absorbance x Energy (photon) graph. Extrapolating the line from the graph by linear regression allows the gap energy of the materials to be determined.

Photodegradation study

The photodegradation tests were carried out using the following procedure: 0.1g of the nanocomposite was dispersed in 15 cm diameter petri dishes and then 15.0 mL of a 1.0×10^{-5} mol L^{-1} aqueous solution of the methylene blue dye was added. The plates were covered with PVC film (Majipack), placed on a tray lined with aluminium foil (to increase the incidence of light) and placed in the sun. The irradiation was carried out between 8:00 and 17:00, as this is the time when UV radiation is highest (source:

Centre for Weather Forecasting and Climate Studies).

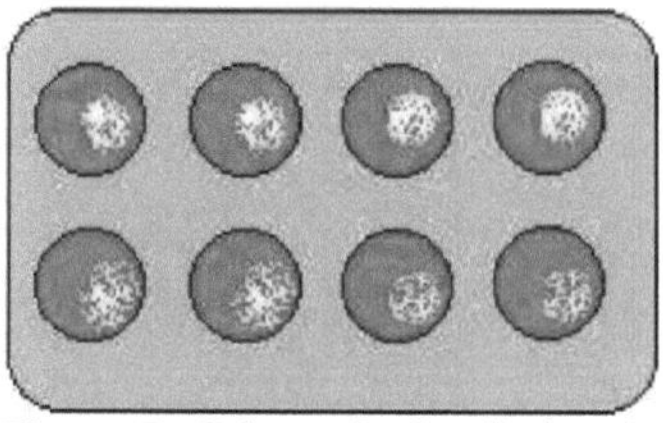

Figure 5 - Scheme for irradiating the samples.

Source: Prepared by the author

The photodegradation of the methylene blue dye in solution was monitored using a Milton-Roy UV/Vis spectrophotometer, model Genesis II, with a scanning system between 200 and 700 nm in a 1.0 cm quartz cuvette.

Contact time study

The study was carried out using ST20, ST30 and ST45 materials. The Petri dishes were prepared as described above and placed in the sun. Irradiation was carried out using one catalyst at a time. The methylene blue solution was kept at its natural pH (5.5).

Every 30 minutes an aliquot of the solution was withdrawn using a syringe to obtain the UV/Vis spectrum.

Study the influence of the solution's pH

To study the influence of the solution's pH on the photodegradation process, several methylene blue solutions at a concentration of $1.0x10^{-5}$ mol L^{-1} were prepared with the pH adjusted from 2 to 8, using NaOH and HCl solutions.

Study of concentration variation

The initial concentration of the dye was varied from $2.0x10^{-6}$ to $3.0x10^{-5}$mol L^{-1}

at natural pH. The solutions used were freshly prepared by dilution from a stock solution with a concentration of $1.0x10^{-3}$mol L^{-1}.

CHAPTER 3

RESULTS AND DISCUSSION

Preparation of Nanocomposites (ST)

The sol-gel process was chosen for the preparation of the nanocomposites due to its peculiar aspects. The sol-gel technique enables a mild temperature reaction, high control of the nanomeric architecture of the material, high dispersion of the forming elements, high surface area, the possibility of forming nanocrystalline structures and a high purity product, as well as producing materials that do not require further purification and do not produce waste to be treated or disposed of.

Characterisation

Scanning electron microscopy

The SEM images and EDS mapping of the particles (Figures 6, 7 and 8) show that the materials have a high dispersion and homogeneity of the forming elements (Si and Ti), since at the magnification used it is not possible to see the presence of phases or agglomerates in the prepared materials. The images suggest that the materials are not formed by surface doping, or a physical mixture of the SiO2 and TiO2 species, which is a first indication that the forming species are intimately mixed (LEE et al., 1993; LI et al., 2010).

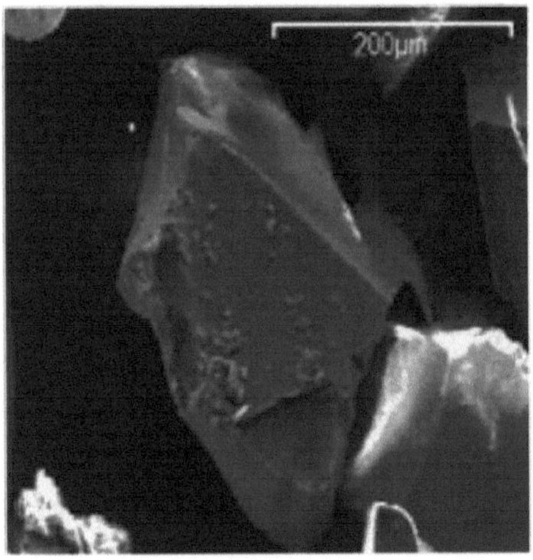

(a)

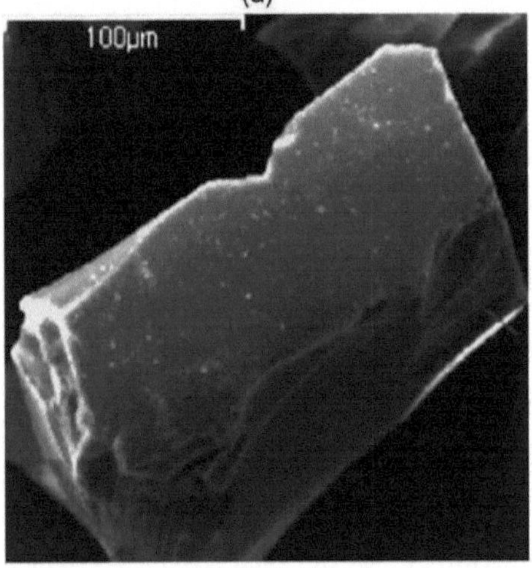

(b)

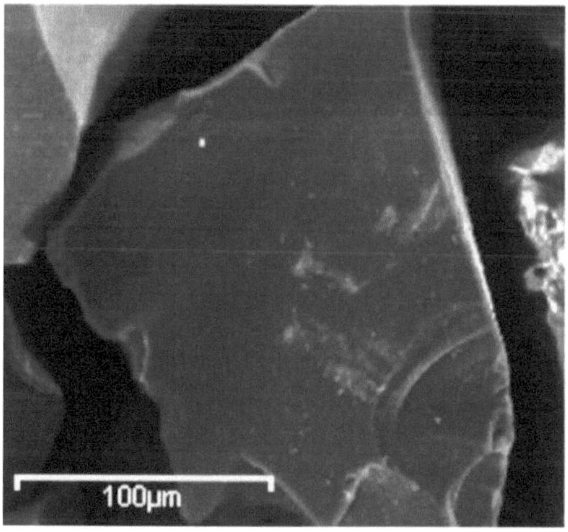

(c)

Figure 6 - SEM images of samples (a) ST45, (b) ST30 and (c) ST20.

Source: Prepared by the author

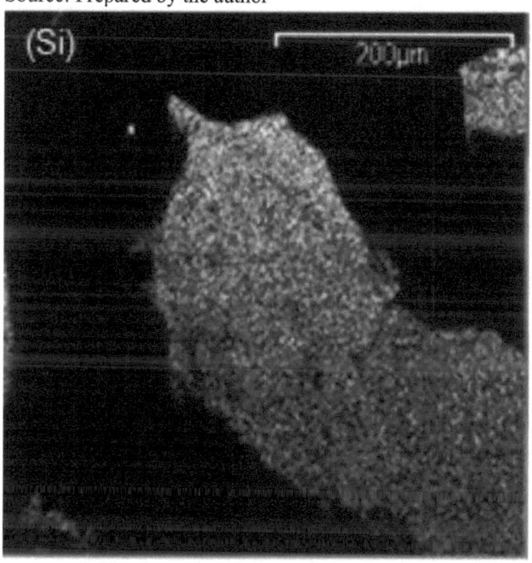

(a)

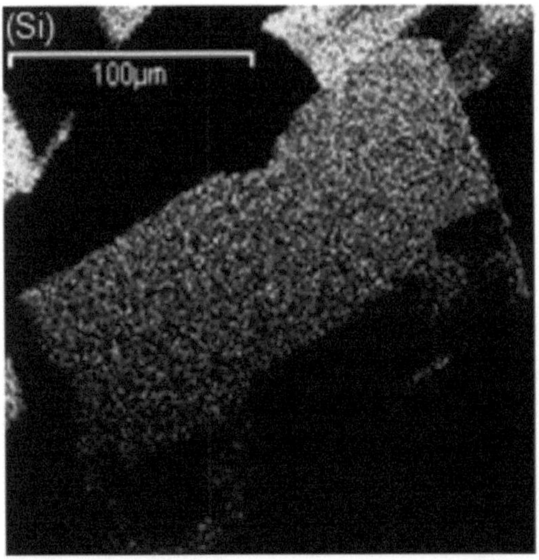

(b)

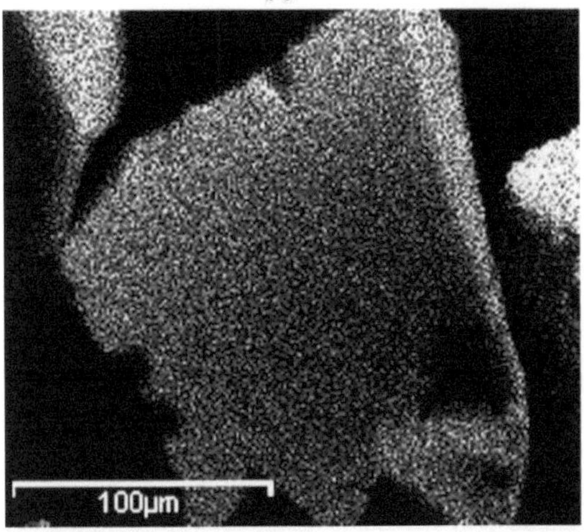

(c)

Figure 7 - EDS mapping of samples (a) ST45, (b) ST30 and (c) ST20.
Source: prepared by the author

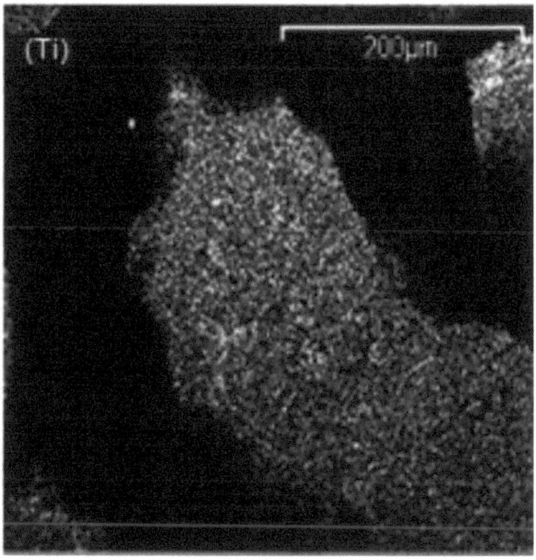

(a)

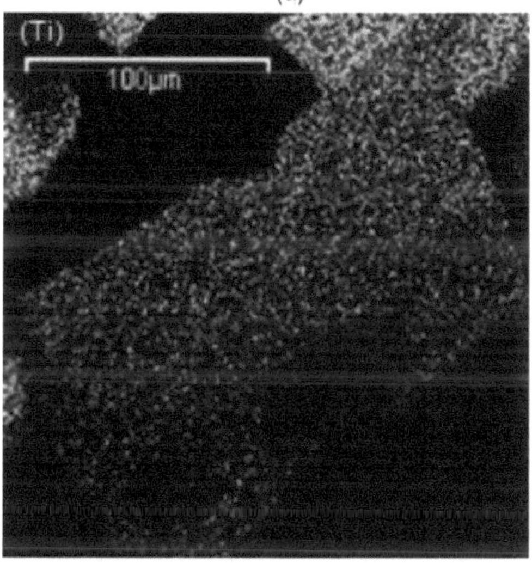

(b)

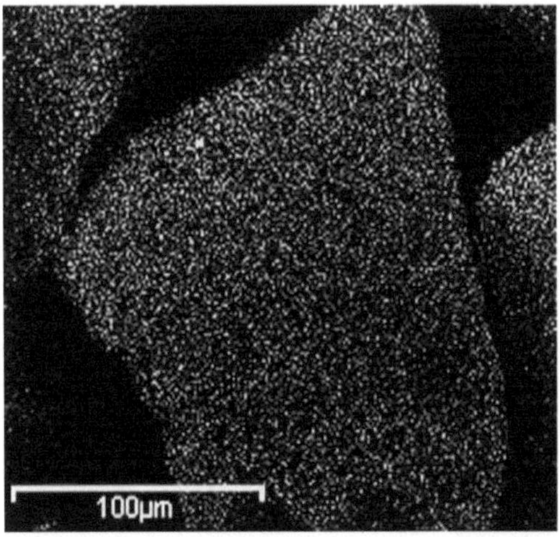

(c)

Figure 8 - EDS mapping of samples (a) ST45, (b) ST30 and (c) ST20.
Source: Prepared by the author

The EDS spectra (Figures 9 to 11) of the material particles showed, in addition to the elements Si and Ti, oxygen (justifiable because they are oxides) and carbon (from the tape used to fix the samples during analysis).

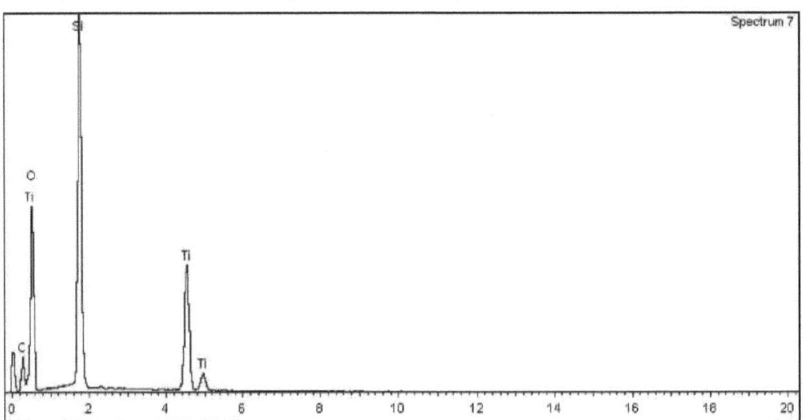

Figure 9 - EDS spectrum of the particles in sample ST20.

Source: Prepared by the author

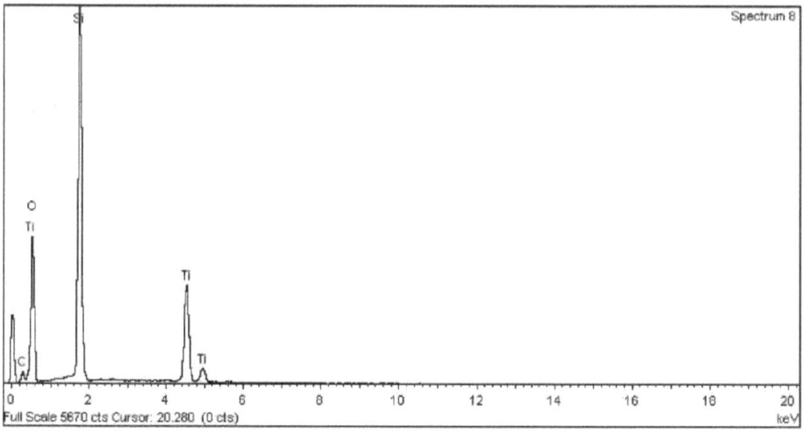

Figure 10 - EDS spectrum of the particles in sample ST30.
Source: prepared by the author

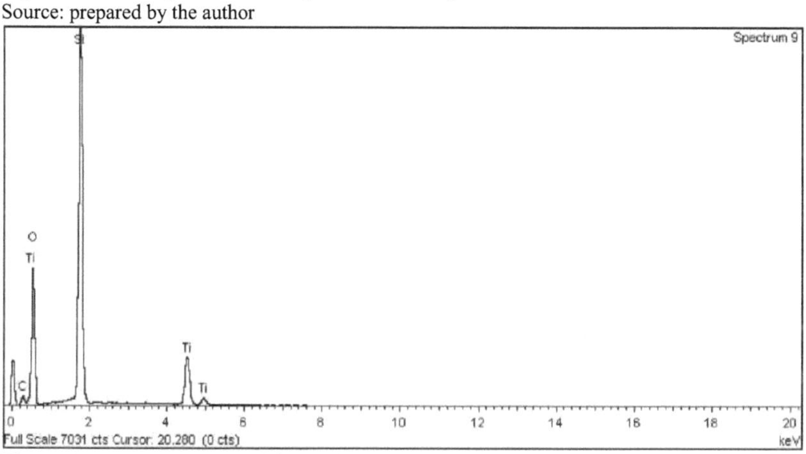

Figure 11 - EDS spectrum of the particles in sample ST45.
Source: Prepared by the author

X-ray diffraction

Powder X-ray diffraction analysis was important for determining possible crystalline phases in the prepared matrices. Figure 12 shows that the materials had no crystalline phases of silica or TiO_2, only the amorphous halo of silica around 25° (2θ/ even after calcination at 500 °C for 120 h to remove all the organic part of the xerogel.

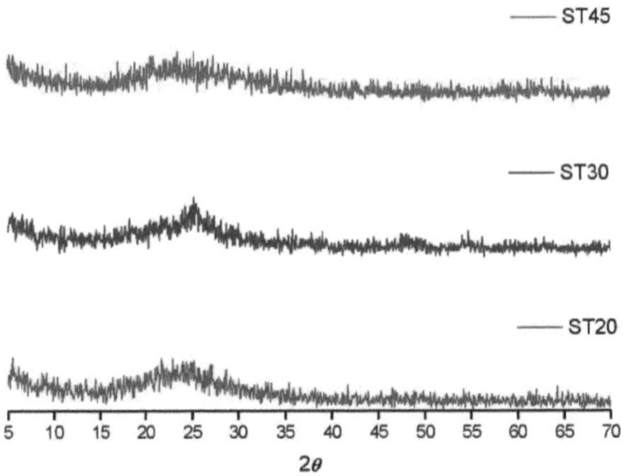

Figura 12 - X-ray diffractograms of the prepared samples.

Source: prepared by the author

FTIR spectrum

The FTIR spectra of the ST20, ST30 and ST45 materials shown in Figure 13 are very similar and the increase in the TiO2 content in the matrices does not promote characteristic changes. The bands at 3430 cm^{-1} and around 1635 cm^{-1} can be attributed to the stretching and angular deformation of the O-H bond of the water adsorbed to the material (LI et al., 2010; YUAN et al., 2010). The broad band with a maximum at approximately 1080 cm^{-1} and the band at 950 cm^{-1} can be attributed to the asymmetric stretching of the Si-O-Si bonds and the Si-OH group bonds present in the matrix (GALINDO et al., 2001). As for the Si-O-Si bonds, symmetrical stretching is also observed, represented by the band around 800 cm^{-1} and angular deformation at approximately 450 cm^{-1} (ZHANG et al., 2010; YUAN et al., 2010).

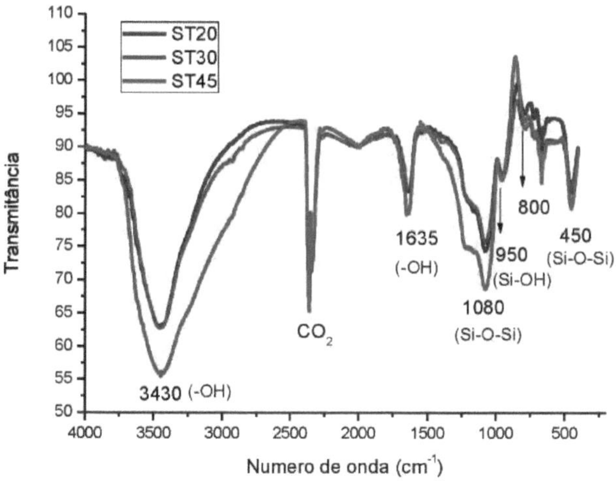

Figura 13 - FTIR spectra of the ST samples.

Source: Prepared by the author

Specific surface area

Analyses of the average surface area and pore volume of the nanostructured materials indicate that the methylene blue molecule is highly accessible to the surface of the matrices. We know that the greater the surface area of a given material, the greater the possibility of this surface interacting with various chemical species. The results of the specific surface areas and average pore volumes shown in Table 4 indicate that the materials are microporous (SATLER, 2011).

Table 4 - Specific surface area and average pore volume of the samples.

Sample	S_{BET} / $m^2\,g^{-1}$	Vp / mL g^{-1}
ST20	765	0,63
ST30	637	0,57
ST45	595	0,52

The data shown in Table 4 shows that as the concentration of titanium in the material increases, the pore volume and surface area of the samples decrease. This behaviour can be explained by the disturbance that occurs in the silica structure with the introduction of TiO_2. Silica is a material made up of regular tetrahedral structures while TiO2 forms octahedral structures (YAN et al., 1995).

Gravimetric analysis

The gravimetric analyses of the samples are shown in Table 5. We can see that there is a great correspondence between the expected SiO2 content and the TiO2 content obtained by chemical analysis. This can be explained by the fact that in sol-gel reactions there are no losses due to washing and purification.

Table 5 - TiO2 and SiO2 content in ST nanocomposites.

Sample	% SiO2	% TiO2
ST20	78,78	21,22
ST30	68,50	31,50
ST45	58,25	41,75

The SiO2 content in the ST materials was determined by difference, i.e. the material was considered to be 100 per cent and the percentage of TiO2 obtained in the chemical analysis was deducted from this value.

Diffuse reflectance

From the diffuse reflectance spectra it was possible to estimate the gap energy of the ST materials and to estimate the size of the nanocrystallites in the nanocomposites. Calculating the gap energy is important for determining whether the materials can be effectively used as photocatalysts. For this to happen, the electron must be able to overcome the energy barrier, generating electron/lacuna pairs. Therefore, the lower the gap energy, the lower the energy required to activate the material and, consequently, the better the photocatalytic potential.

The gap energies were calculated by extrapolating the line from the absorption spectrum curve (Figure 14). The energy values were calculated from the slopes observed in the spectrum of each sample. From the results shown in Table 6, it can be seen that the ST45 composite has the lowest gap energy and therefore the best potential to act as a photocatalyst.

Figures 14 and 15 refer to the ST45 composite, but the other materials were treated in a similar way, thus obtaining the gap energies and crystallite sizes present in the materials.

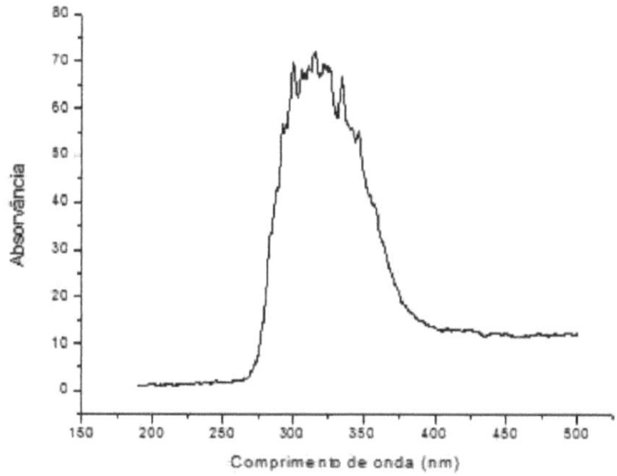

Figure 14 - Diffuse reflectance spectrum for material ST45.

Source: Prepared by the author

By interpolating the linear part of the Tauc graph and considering that the general equation of a straight line is that described by Equation 7

$$Y = a(X) + b \quad \text{(Equação 7)}$$

we can assume that when Y= 0, we will have Equation 8:

$$X = \frac{-b}{a} \quad \text{(Equação 8)}$$

where X is the value of the gap energy (eV), **b** is the value of the intersection of the line on the y-axis (linear coefficient of the line) and **a** is the angular coefficient of the line (slope of the line).

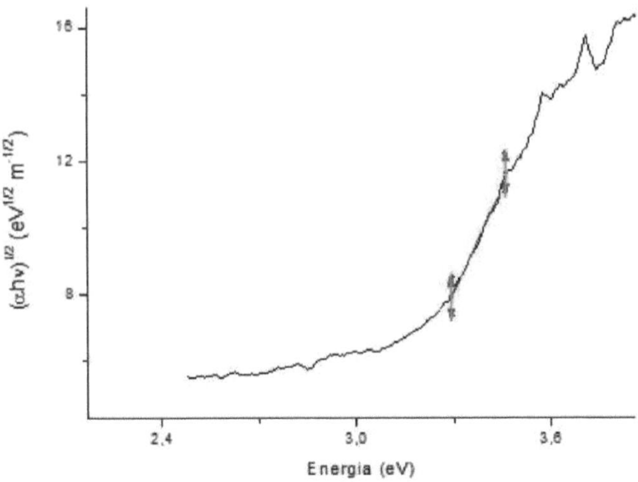

Figure 15 - Tauc graph for the ST45 nanocomposite.
Source: Prepared by the author

Table 6 - Gap energy and disorder factor values for ST samples.

Sample	Disorder factor ($B^{1/2}$) (eV^{-1} - *cm-1*)$^{1/2}$	Gap energy (e/V)
ST20	32,3	3,19
ST30	21,5	2,97
ST45	20,9	2,91

The $B^{1/2}$ parameter indicates the effect of disorder at the edge of the band, so that the greater the disorder, the lower the value of the coefficient (MORIGAKI, 1994). This parameter is essentially due to the contribution of the atoms on the surface of the nanoparticles and is obtained from the value of the slope of the line. It can be seen that the higher the TiO2 content in the system, the more disturbed the band edge.

An exciton is a bound state of an electron and a gap, which are attracted by electrostatic Coulomb forces. The quantum confinement of the exciton means that the continuous energy band acquires a molecular character, with narrow energy bands. The absorption band for direct transitions of nanostructured semiconductors is shifted to higher energies than in the source materials. The photoinduction of high concentrations of electrons in a narrow conduction band with low density of states also changes the gap energy.

It can be seen that the gap energy decreases with increasing TiO2 content in the nanocomposite, which may indicate that the quantum confinement effect is decreasing. In addition, there is also an increase in the structuring of a TiO_2 phase or phases, which could explain the decrease in the band disorder coefficient (STUCKY; DOUGALL, 1990).

The effective mass model was applied to determine the size of the crystallites.

If one charge carrier approaches another, the formation of a bound state (Wannier exciton) can occur, which can be approximated by the hydrogenionic Hamiltonian in equation 9 (BRUS, 1984):

$$\hat{H} = -\frac{\hbar}{2m_h}\nabla_h^2 - -\frac{\hbar}{2m_e}\nabla_e^2 - \frac{e^2}{\varepsilon|r_e - r_h|} \quad \text{(Equação 9)}$$

The solution to equation 9 for the lowest energy excited state is shown in equation 10.

$$E_g = E_{g,C} + \frac{\hbar^2\pi^2}{2r^2}\left(\frac{1}{m_e^*} + \frac{1}{m_h^*}\right) - \frac{1.8e^2}{\varepsilon r} + \sum_{n=1}^{\infty}\frac{e^2}{r}\alpha_n\left(\frac{r_e + r_h}{r}\right)^{2n} \quad \text{(Equação 10)}$$

Where $E_{g.C}$ is the band-gap value of the continuous phase of the sample, the second term describes the kinetic energy increment in the localisation of the electron/lacunar pair inside a sphere with radius R (where h^2 is Planck's constant squared divided by π^2) and m_h*, m_e* are the effective mass of the gap and the electron respectively. The third term in Equation 10 indicates the Coulombian attraction for the selected region, where e^2 is the electron charge squared and ε is the permittivity in the medium. It is then possible to simplify Equation 10 to a reduced form represented by Equation 11, where we have the presence of constant terms (JACOBSSON; EDVINSSON, 2011):

$$E_g = C_1 + \frac{C_2}{r} + \frac{C_3}{r^2} \quad \text{(Equação 11)}$$

For the system under study, the following approximations were applied to the constants C_1, C2 and C_3:

- Approach 1:

It was considered:

c_1= 3.2 (eV) (SILVA et al., 2010);

c_2= - 8.57(eV nm) (MUKHERJEE et al., 1994) and

c_3= 11.74(eV nm^2) (SERPONE et al., 1995).

Estimate for lighter excitons.

- Approach 2:

It was considered:

c_1= 3.2(eV) (SILVA et al., 2010);

c_2= - 8.57 (eV nm) (MUKHERJEE et al., 1994) and

c_3= 30.51(eV nm^2) (SERPONE et al., 1995).

Estimate for heavier excitons

In other words, a system where the lowest energy excited state corresponds to TiO_2.

Figure 16 shows the curves obtained from each approximation. The crystallite size is obtained by interpolating the linear part of the graph with the X axis.

The approximations consider TiO2 as a charge carrier generator. It should be borne in mind that the effective mass model overestimates the kinetic energy of excitons (the term that represents the c_3 constant), so it can be observed that the values found may underestimate the real crystallite sizes.

Table 7 shows the results estimated by the mathematical approximations for the crystallite sizes present in the ST materials.

Table 7 - Crystallite size using the effective mass approximation.

Sample	Approach 1 (nm)	Approach 2 (nm)

ST20	1,34	3,57
ST30	1,39	3,99
ST45	1,41	4,14

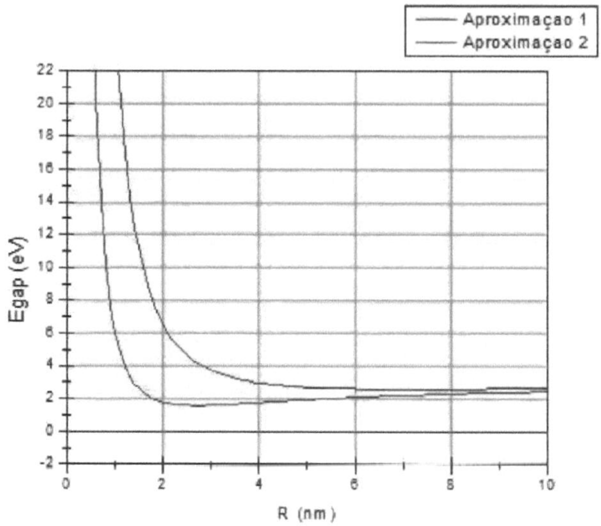

Figure 16 - Energy versus crystallite size using the effective mass approximation.

Source: Prepared by the author

Photodegradation Studies with Sunlight

Photodegradation studies of the methylene blue dye in aqueous solution without pH adjustment were carried out in natural sunlight using the ST20, ST30 and ST45 nanocomposites. These preliminary studies of the contact time behaviour of the photocatalysts with the dye solution were carried out with the aim of determining which nanocomposite is the most efficient in photodegradation with sunlight.

The photodegradation process of the dye was monitored using UV/Vis analyses as a function of irradiation time. In order to make it easier to monitor the photodegradation process, the wavelengths 292.11 nm (aromatic region) and 615.58 and 663.69 nm (chromophore group region) were chosen as control wavelengths.

The aqueous solution of methylene blue showed maximum absorbance at 670nm, which is attributed to the monomeric species of the dye (REIS; SERRANO; PASTRE).

Influence of contact time

Figures 17, 18 and 19 show the photodegradation behaviour of the methylene blue dye in relation to the contact time for each type of nanocomposite. By observing the data in Figures 17, 18 and 19 we can see that catalytic degradation is very fast in water, with a maximum exposure time of 2 hours being necessary for the methylene blue dye at a concentration of $1.0x10^{-5}$ mol L^{-1} to be completely mineralised.

Considering the maximum absorbance of methylene blue at 670 nm and the molar absorptivity coefficient of the dye in water, the Lambert-Beer Law (Equation 12) can be used to determine the concentration of the dye in the solution at a given irradiation time.

$$A = \varepsilon bC \qquad \text{(Equação 12)}$$

A represents the absorbance, ε the molar absorptivity coefficient of methylene blue (71089 mol^{-1} cm^{-1} L), b is the optical path (1.0 cm) and C is the concentration of methylene blue in the solution in mol L^{-1}.

Table 8 shows the photodegradation rate for the $1.0x10^{(-5)}$mol L^{-1} solution, considering the different photocatalysts used.

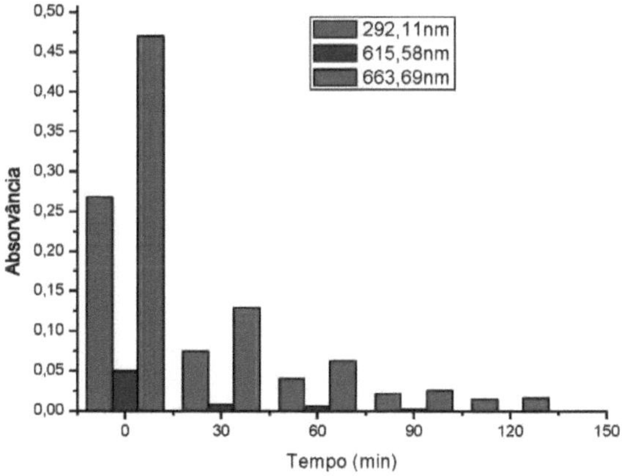

Figura 17 Behaviour of methylene blue photodegradation with ST20 as a function of irradiation time.

Source: Prepared by the author

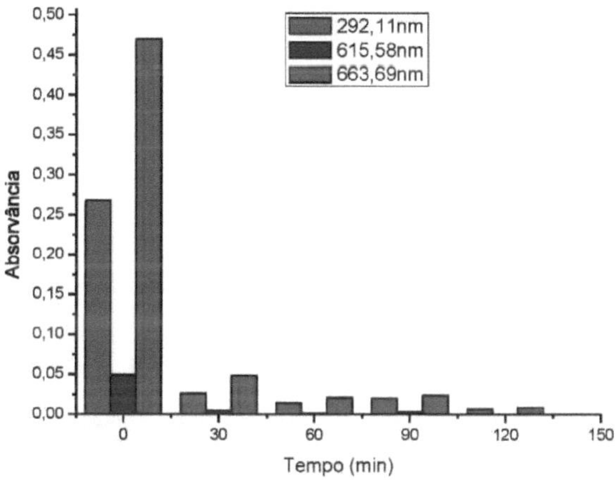

Figura 18 - Behaviour of methylene blue photodegradation with ST30 as a function of irradiation time.

Source: Prepared by the author

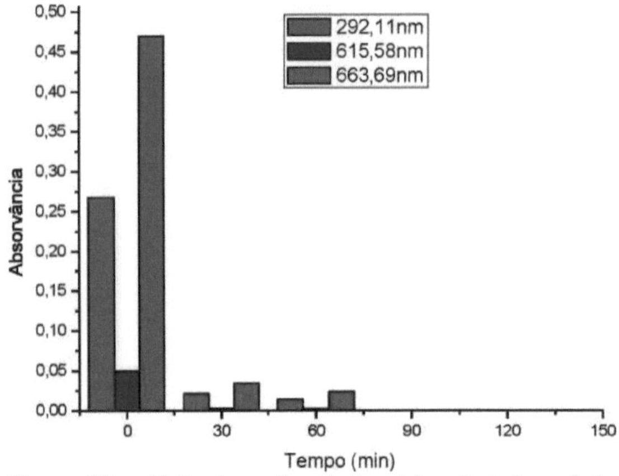

Figura 19 - Behaviour of methylene blue photodegradation with ST45 as a function of irradiation time.

Source: Prepared by the author

Table 8 - Methylene blue concentration values at time zero, after 30 minutes of irradiation and photodegradation rate.

Sample	C0 (mol L^{-1})	C_{30min} (mol L^{-1})	Photodegradation rate (%)
ST20	1,0x10-5	1,8x10-6	82,0
ST30	$1{,}0x10^{-5}$	$6{,}9x10^{-7}$	93,1
ST45	$1{,}0x10^{-5}$	$4{,}8x10^{-7}$	95,2

Considering the initial concentrations and after 30 minutes of irradiation, it can be seen that there is a 95.2% reduction in relation to the initial concentration when the ST45 catalyst is used. This result shows that the greatest efficiency in the degradation of the dye occurs with the use of ST45, a result consistent with the study of gap energies, which is the most significant parameter for the efficiency of photocatalysts.

Influence of the solution's pH

The other studies were carried out only with the ST45 nanocomposite, since it showed the best photodegradation efficiency compared to the other nanocomposites prepared.

In order to determine the influence of the solution's pH on the photodegradation process, it is necessary to determine whether the methylene blue molecule in aqueous solution undergoes significant changes in its UV/Vis spectrum. Therefore, the UV/Vis spectrum of methylene blue solutions at a concentration of $1.0x10^{-5}$ mol L^{-1} with pH adjusted to different values was scanned to detect possible changes in the electronic spectrum and, therefore, in the dye molecule. The spectra obtained are shown in Figure 20.

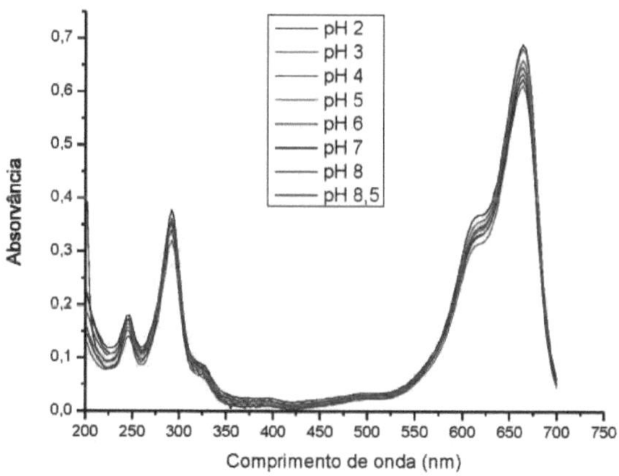

Figura 20 - Electronic spectra of the $1.0x10^{-5}$ mol L^{-1} solution of methylene blue at different pH values.

Source: Prepared by the author

In this case, no variations were observed in the electronic spectrum of the different solutions, such as an increase or decrease in absorbance or a shift in the absorption bands. It can therefore be concluded that there is no change in the structure of the dye in the pH range used in the photodegradation study.

Once it has been confirmed that the methylene blue molecule does not undergo significant changes in its UV/Vis spectrum, it is possible to begin studying the photocatalytic behaviour of ST45 at various pH values adjusted for the $1.0x10^{-5}$ mol L^{-1} methylene blue solutions. Figures 21 to 28 show the photodegradation behaviour of methylene blue at different pH values, using nanocomposite ST45 and natural sunlight.

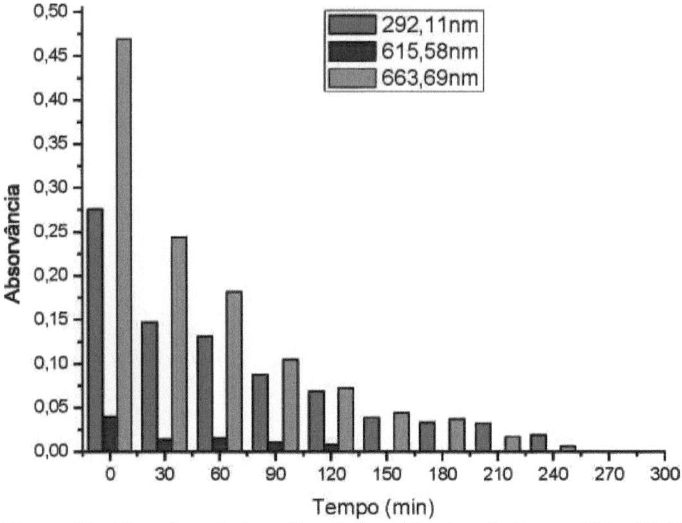

Figure 21 - Photodegradation of methylene blue in solution at pH 2 with ST45.

Source: Prepared by the author

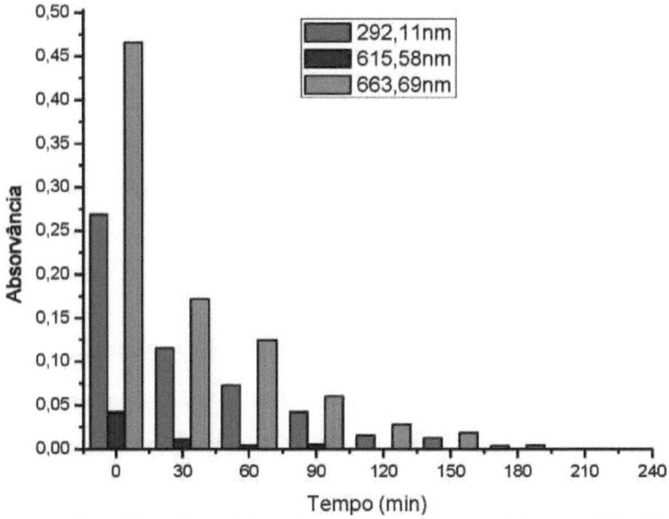

Figura 22 - Photodegradation of methylene blue in solution at pH 3 with ST45.
Source: Prepared by the author

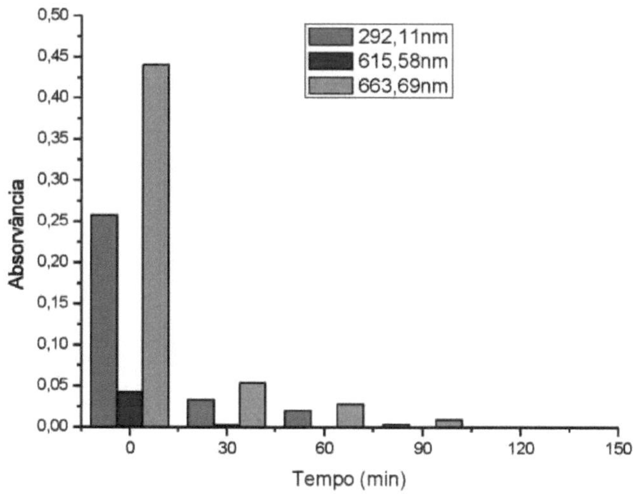

Figura 23 - Photodegradation of methylene blue in solution at pH 4 with ST45.
Source: Prepared by the author

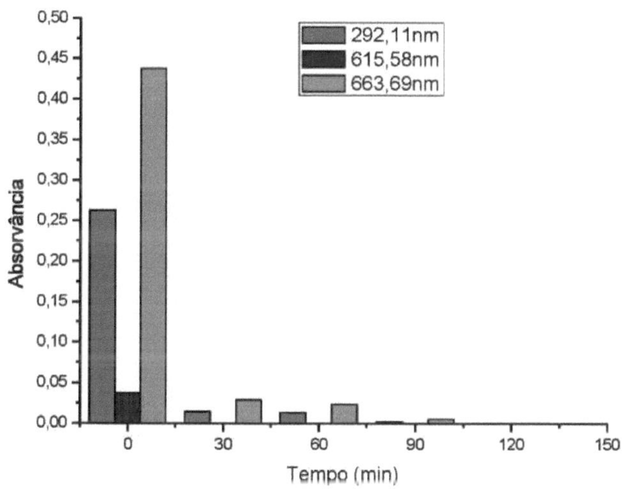

Figura 24 - Photodegradation of methylene blue in solution at pH 5 with ST45.

Source: Prepared by the author

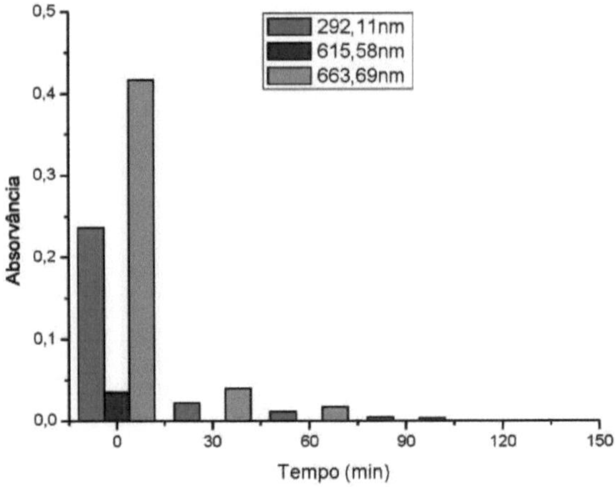

Figura 25 - Photodegradation of methylene blue in solution at pH 6 with ST45.

Source: Prepared by the author

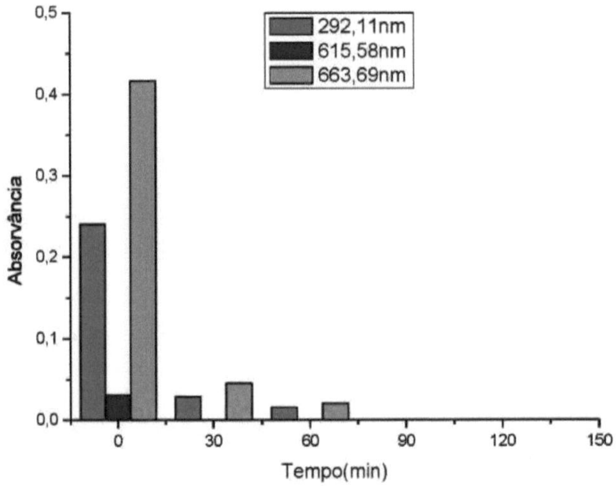

Figura 26 - Photodegradation of methylene blue in solution at pH 7 using ST45.

Source: Prepared by the author

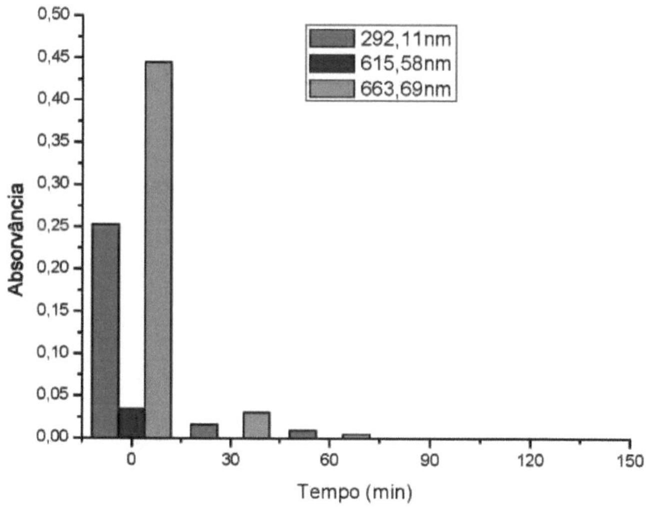

Figura 27 - Photodegradation of methylene blue in solution at pH 8 with ST45.

Source: Prepared by the author

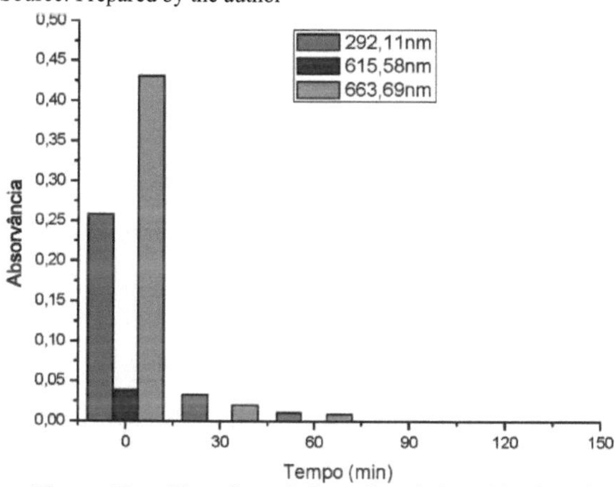

Figura 28 - Photodegradation of methylene blue in solution at pH 8.5 with ST45.

Source: Prepared by the author

By observing the photodegradation behaviour of methylene blue, shown in Figures 21 to 28, it can be seen that at very low pH values (2 and 3), the photodegradation process is slower. This fact may be associated with strong competition between the protons in solution and the methylene blue molecules for the surface of the photocatalyst, highlighting the fact that real adsorption of the organic molecule on the surface of the photocatalyst is necessary for the degradation process to take place efficiently.

The pH values of the solutions at the end of the irradiation time were determined and it was clear that the initial pH 2 to 4 solutions did not undergo a significant change in value. However, the solutions at pH 6 to 8.5 showed a decrease in pH value, taking on values between 4 and 5 (Figure 29).

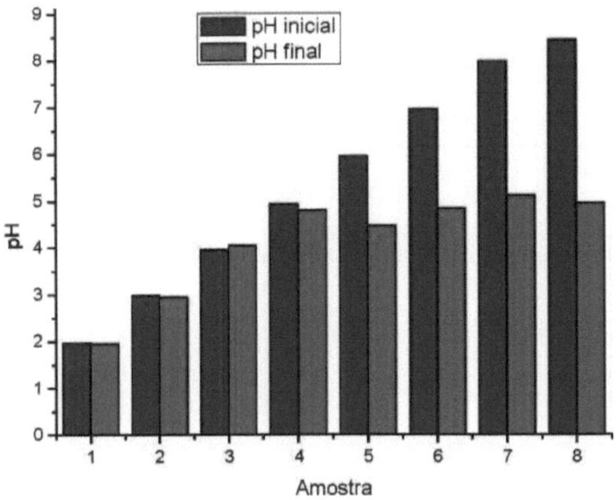

Figura 29 - Variation in the pH of methylene blue solutions in the photodegradation process.

Source: Prepared by the author

From the analysis of the photodegradation behaviour of methylene blue shown in Figure 29, it is assumed that, at very low pH values, there is a large amount of protons in the solution that remain throughout the photodegradation process. However, at higher values, there seems to be a more or less constant final pH value in the 5.0 range. This behaviour suggests that total mineralisation of the system is taking place and, most likely, in addition to CO_2 and H_2O, there is also the production of ammonium sulphate in the aqueous solution.

Influence of concentration variation

Varying the concentration of the methylene blue solution may indicate an interesting dye/photocatalyst ratio in terms of application potential. Figures 30 to 33 show the photodegradation behaviour as a function of solar irradiation time when the amount of photocatalyst is fixed (0.1 g) and different concentrations of the dye are used.

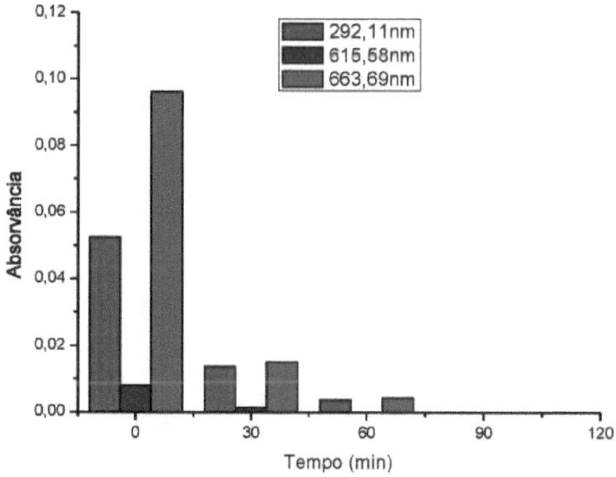

Figura 30 - Photodegradation of methylene blue using 0.1 g of ST45 and a solution of $2.0x10^{-6}$mol L^{-1}.

Source: Prepared by the author

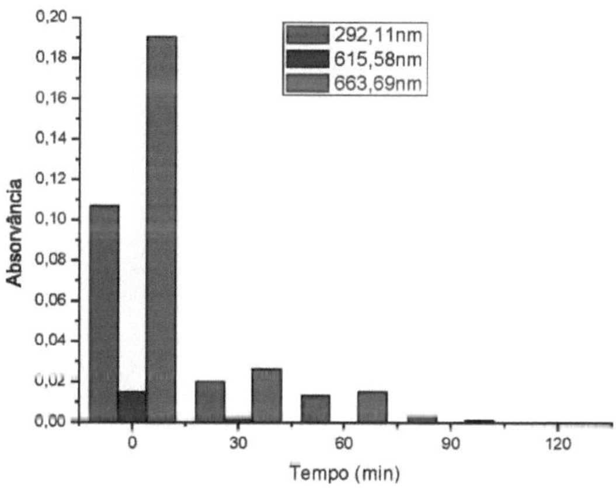

Figura 31 - Photodegradation of methylene blue using 0.1 g of ST45 and a solution of $4.0x10^{-6}$mol L^{-1}.

Source: Prepared by the author

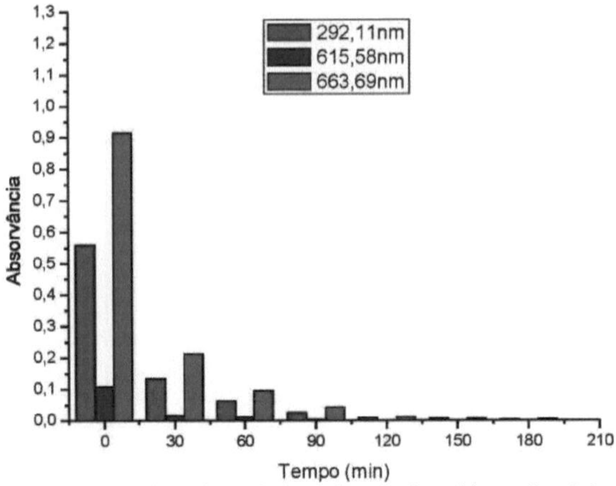

Figura 32 - Photodegradation of methylene blue using 0.1 g of ST45 and a solution of $2.0x10^{-5}$ mol L^{-1}.

Source: Prepared by the author

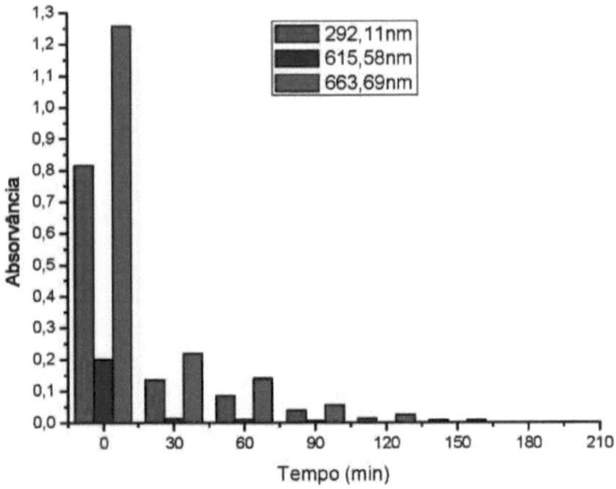

Figura 33 - Photodegradation of methylene blue using 0.1 g of ST45 and a solution of $3.0x10^{-5}$ mol L^{-1}.

Source: Prepared by the author

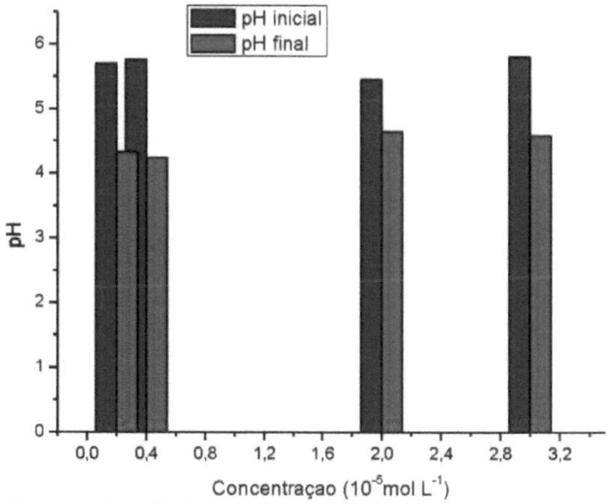

Figure 34 - pH of the initial methylene blue solution and at the end of photodegradation

Source: Prepared by the author

Even at higher concentrations, the ST45 photocatalyst showed great efficiency in degrading the dye, requiring only 3 hours of solar irradiation for complete degradation.

With regard to the pH behaviour of the solution, shown in Figure 34, it can be seen that there is probably the final formation of a salt, which could be ammonium sulphate.

Photocatalyst recovery

The photocatalysts used in the various experiments were recovered as follows: the solid materials were washed after the experiment and kept in a beaker with Milli-Q water. When a certain amount had been collected, the mixture was placed in the sun for 6 hours. At the end of the process, the solid material was filtered, washed and dried at 25°C for the next use.

It is estimated from the number of experiments carried out and their various repetitions that the nanocomposites were recycled at least 350 times over a period of 18 months.

With the real possibility of recovering the photocatalysts, it is assumed that the matrix is rigid and the TiO_2 does not leach out in relation to the time of use and the pH values of the solutions used. This behaviour shows that the nanocomposites prepared in this work are versatile and stable under the conditions used.

The waste from the methylene blue dye solutions is being treated by ST45 and sunlight in order to achieve a minimum volume of waste in the project.

CHAPTER 4

FINAL CONSIDERATIONS

The sol-gel process is an easy and efficient method for preparing nanocomposites without generating waste.

Characterisation of the ST materials showed that the matrices have a high specific surface area with a small average pore volume and are therefore microporous; they have no crystalline phases identifiable by powder X-ray diffraction; they have a high homogeneity in the dispersion of the elements throughout the material.

With regard to photodegradation studies of methylene blue dye in aqueous solution, using sunlight and ST nanocomposites, the materials showed excellent potential for photodegradation.

The ST materials used in this work were recycled more than 350 times over a period of 18 months, with no apparent reduction in their efficiency.

The photodegradation tests did not use oxidising agents such as peroxides to increase the efficiency of the photocatalysis. The hydroxyl radicals are generated from the water used to prepare the solutions. In this way, the methodology presented in this work is low cost and does not generate waste that has a high impact on the environment.

The possibility of using only sunlight as an energy source has significant value, as it is an environmentally friendly and economically sustainable technology.

BIBLIOGRAPHICAL REFERENCES

BANSAL, P., SUD, D. Photodegradation of commercial dye, Procion Blue HERD from real textile wastewater using nanocatalysts. **Desalination.** 267, 2011, 244-249

BIANCO-PREVOT, A. BAIOCCHI, C., BRUSSINO, M.C., PRAMAURO, E., SAVARINO, P., AUGUGLIARO, V., MARCI, G., PALMISANO, L. Photocatalytic degradation of acid blue 80 in aqueous solutions containing TiO2 suspensions. **Environ Sci Technol.** 35, 2001, 971-976.

ESTEVES, A. C. C.; BARROS-TIMMONS, A.; TRINDADE, T. Polymer matrix nanocomposites: strategies for synthesising hybrid materials. **Química Nova.** 27, 5,

2004, 798-806.

GALINDO, C., JACQUES, P., KALT, A. Photochemical and photocatalytic degradation of an indigoid dye: a case study of acid blue 74 (AB74). **J Photoch Photobio A.** 141,2001,47-56.

HERNÁNDEZ, A., MAYA, L. SÁNCHEZ-MORA, E., SÁNCHEZ, E. M. Sol-gel synthesis, characterisation and photocatalytic activity of mixed oxide ZnO- Fe_2O_3. **J Sol-Gel Sci Techn.** 42, 2007, 71-78.

HOUAS, A., LACHHEB, H., KSIBI, M., ELALOUI, E., GUILLARD, C., HERMANN, J.M. Photocatalytic degradation pathway of methylene blue in water, **Appl Catal B-Environ.** 31, 2001, 45-157.

JACOBSSON, J.; EDVINSSON, T. Absorption and Fluorescence Spectroscopy of Growing ZnO Quantum Dots: Size and Band-gap Correlation and Evidence of Mobile Trap States. **Inorg Chem.** 50, 2011, 9578-9586.

KUNZ, A.; PERALTA-ZAMORA, P.; MORAES, S. G.; DURÁN. N. New trends in the treatment of textile effluents. **Química Nova**. 25, 1, 2002, 7882.

LEE, R. E. **Scanning Electron Microscopy and X-Ray Microanalysis.** New Jersey: Prentice Hall, 1993, 458 p. 168-406.

LI, G.; LIU, F; ZHANG, Z. Enhanced photocatalytic activity of silica-embedded TiO2 hollow microspheres prepared by one-pot approach. **J. Alloy Compd.** 493, 2010, L1-L7.

MORIGAKI, K. **Physics of Amorphous Semiconductors**. Singapore: World Scientific. 1999.

MUKHERJEE, N.; MYERS, R. A.; BRUECK, S.R.J. Dynamics of second- harmonic generation in fused silica. **J Opt Soc Am B.** 11, 1994, 655-668.

MURPHY, A.B. Band-gap determination from diffuse reflectance measurements of semiconductor films and application to photoelectrochemical water-splitting. **Sol Energ Mat Sol C**. 91, 2007, 1326-1337.

NOGUEIRA, R. F. P.; JARDIM, W. F. Heterogeneous photocatalysis and its environmental application. **Química nova**. 21,1, 1998, 69-71.

NUNES, A. P. B. **Synthesis, characterisation and applications of nanostructured TiO2 photocatalysts**. 2011. 81 f. Dissertation (Master's in Materials Science) - Federal University of Rio Grande do Sul, Porto Alegre.

OZCAN, A.; ONCU, A.S. Kinetics, isotherm and thermodynamic subtides of adsorption of Acid Blue 193 from aqueous solution onto natural sepiolite. **Colloid Surfaces A.** 277, 2006, 90-97.

PEREIRA, A. C.; KUBOTA, L. T.; SANTOS, A. S. Trends in modification of amperometric electrodes for electroanalytical applications. **Química Nova.** 25, 6, 2002, 1012-1021.

REIS, F. A., SERRANO, L., CAPORALIN, C. B., PASTRE, I. A. Spectrophotometric study of methylene blue dye in modified clays. Available at: <http://prope.unesp.br/xxi_cic/27_35607566854.pdf> Accessed on: 07 Oct. 2012

ROBINSON, T.; MCMULLAN, G.; MARCHANT, R.; NIGAM, P. Remediation of dyes in textile effluent: a critical review on current treatment technologies with a proposed alternative. **Bioresource Technol.** 77, 2001, 247-255.

SATTLER, K. D. (Ed.) **Functional Nanomaterials.** Handbook of nanophysics 5: Boca Raton: CRC press, 2011.

SERPONE, N.; LAWLESS, D.; KHAIRUTDINOV, R. Size Effects on the Photophysical Properties of Colloidal Anatase TiO2 Particles: Size Quantisation or Direct Transitions in This Indirect Semiconductor? **J. Phys. Chem.** 99,1995, 16646-16654.

SILVA, S. S.; MAGALHÃES, F.; SANSIVIERO, M. T. C. Semiconductor nanocomposites of ZnO/TiO2 photocatalytic studies. **Química Nova.** 33, 2010, 85-89

SOARES J.L. **Removal of textile dyes by adsorption on activated mineral carbon with a high ash content**. 1998. 99 f. Dissertation (Master's Degree in Chemical Engineering) - Federal University of Santa Catarina, Florianópolis.

STENGL, V.; KRÁLOVÁ, D. TiO2/ZnS/CdS Nanocomposite for Hydrogen Evolution and Orange II Dye Degradation. **Int J Photoenergy**. 2011.

STUCKY, G.D., DOUGALL, J.G.M. Quantum confinement and Host/Guest chemistry: probing a new dimension. **Science.** 247, 1990, 669-678.

STRAUSS , M.; MARONEZE, C. M.; SILVA, J. M. S; SIGOLI, F. A.; GUSHIKEM, Y.; MAZALI, I. O. Annealing temperature effects on sol-gel nanostructured mesoporous SiO2/TiO2 and its photocatalytic activity. **Mat Chem Phys**. 126, 2011,

188-194.

Tabulated Molar Extinction Coefficient for Methylene Blue in Water. Available at: <http://omlc.ogi.edu/spectra/mb/mb-water.html> Accessed on: 07 Oct. 2012

YAN, J.; BUCKLEY, A. M.; GREEBLATT, M. The preparation and characterisation of silica gels doped with copper complexes. **J Non-Cryst Solids.** 180, 1995, 180-190.

YUAN, Q.; LI, N.; JINCHUN, T.; LI, X.; WANG, R.; ZHANG, T.; SHAO, C. Preparation and humidity sensitive property of mesoporous ZnO-SiO2 composite. **Sensors Actuat B-Chem** 149, 2010, 413-419.

ZHANG, L.; LI, J.; CHEN, Z.; TANG, Y.; YU, Y. Preparation of Fenton reagent with H2O2 generated by solar light-illuminated nano-Cu2O/MWNTs composites. **Appl Catal. A**. 299, 2006, 292-297.

ZANONI, M. V. B., CARNEIRO, P.A. O descarte dos colorantes têxteis.**Ciência Hoje.** v. 29, n. 174,61 - 64,agosto, 2001.

MIX
Papier aus verantwortungsvollen Quellen
Paper from responsible sources
FSC® C105338

Printed by Books on Demand GmbH, Norderstedt / Germany